SHARK FILES

BULL SHARKS

by Julie K. Lundgren

A Crabtree Seedlings Book

TABLE OF CONTENTS

Crabtree Publishing
crabtreebooks.com

School-to-Home Support for Caregivers and Teachers

This book helps children grow by letting them practice reading. Here are a few guiding questions to help the reader with building his or her comprehension skills. Possible answers appear here in red.

Before Reading:

- What do I think this book is about?
    - *I think this book is about bull sharks.*
    - *I think bull sharks are very large.*
- What do I want to learn about this topic?
    - *I want to learn why sharks bite people.*
    - *I want to learn the size of bull sharks.*

During Reading:

- I wonder why...
    - *I wonder why bull sharks hunt day and night.*
    - *I wonder why remoras keep bull sharks clean.*
- What have I learned so far?
    - *I have learned that bull sharks can live in saltwater and freshwater.*
    - *I have learned that bull sharks mostly live in saltwater near land.*

After Reading:

- What details did I learn about this topic?
    - *I have learned that freshwater rivers and lakes are not salty like oceans.*
    - *I have learned that young bull sharks are called pups.*
- Read the book again and look for the vocabulary words.
    - *I see the word **remoras** on page 16 and the word **stout** on page 8. The other glossary words are on pages 22 and 23.*

BULL SHARKS

Chomp! Bite! Snap!

Hungry bull sharks
hunt day and night.

Bull sharks eat fish and **rays**.

FROM THE FILES

Sharks have rows of sharp **teeth**.

Bull sharks are **stout** like bulls.

FROM THE FILES

Their young are called pups.

Bull sharks weigh between 200 and 500 pounds (90 and 227 kg).

They live in **saltwater** near land.

Unlike most other sharks, they can also live in **freshwater**.

Remoras keep bull sharks clean.

Remoras eat food bits stuck to the shark's body.

After their cleaning, bull sharks return to the hunt.

GLOSSARY

freshwater (FRESH-wah-tur): Freshwater rivers and lakes are not salty like oceans.

rays (RAYZ): Rays are ocean animals with wide, flapping fins for swimming, and long, thin tails.

remoras (ruh-MOR-uhz): Remoras are fish that clean leftover food bits stuck to sharks.

saltwater (SAHLT-wah-tur): Saltwater is the salty water that fills Earth's oceans.

stout (STOUT): Stout animals have thick, wide bodies.

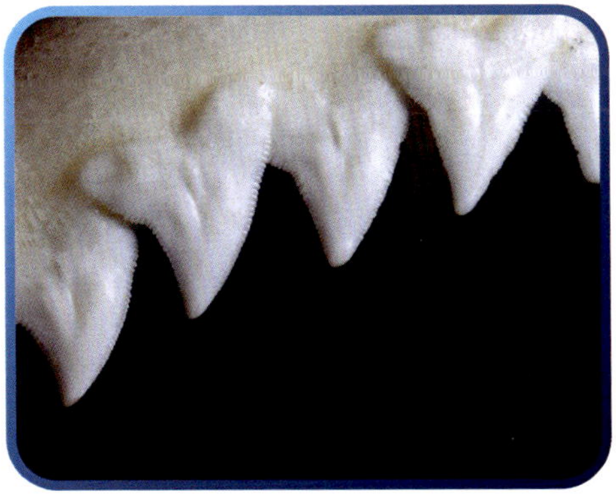

teeth (TEETH): Teeth are white, bony parts inside a mouth that are used for biting and chewing.

Index

Websites

https://kids.nationalgeographic.com/animals/fish/bull-shark

https://easyscienceforkids.com/all-about-sharks

About the Author

Julie K. Lundgren

Julie K. Lundgren grew up near Lake Superior where she reveled in mucking about in the woods, picking berries, and expanding her rock collection. Her interests led her to a degree in biology. She lives in Minnesota with her family.

Crabtree Publishing

crabtreebooks.com 800-387-7650

Written by: Julie K. Lundgren
Designed by: Jennifer Dydyk
Edited by: Kelli Hicks
Proofreader: Melissa Boyce

Photographs:
Shark illustration on cover logo © BATKA/Shutterstock; white shark illustration for "FROM THE FILES" © Dashikka/Shutterstock; Cover photo © Andrea Izzotti/Shutterstock; page 3 © Carlos Grillo/istock; page 5 (top) © Divepic/istock, (bottom) © izanbar/istock; page 6 © Matthew R McClure/Shutterstock; page 7 © June Jacobsen/istock; page 9 (photo) © 97420280/Shutterstock, (illustration) © Bullet_Chained/istock; page 11 (tractor) © Valentin Valkov/Shutterstock, (shark) © Divepic/istock; page 13 © Leonardo Gonzalez/Shutterstock; page 14 (photo only) © Valerijs Novickis/Shutterstock; page 15 © dangarneau/istock; page 17 © Fiona Ayerst/Shutterstock; page 19 © FionaAyerst/istock; page 21 © June Jacobsen/istock; page 22 (top) © Doucefleur/istock, (middle) © Shahar Shabtai/Shutterstock; page 23 (top) © Anastasia Deriy/istock

Hardcover	978-1-4271-5830-7
Paperback	978-1-4271-5831-4
Ebook (pdf)	978-1-4271-5832-1
Epub	978-1-4271-5833-8
Read-along	978-1-4271-5834-5
Audio book	978-1-4271-5835-2

Printed in Canada/042024/CP20240422

Published in Canada
Crabtree Publishing
616 Welland Avenue
St. Catharines, Ontario
L2M 5V6

Published in the United States
Crabtree Publishing
347 Fifth Avenue
Suite 1402-145
New York, NY 10016

Library and Archives Canada Cataloguing in Publication

Available at the Library and Archives Canada

Library of Congress Cataloging-in-Publication Data

Names: Lundgren, Julie K., author.
Title: Bull sharks / by Julie K. Lundgren.
Description: New York : Crabtree Publishing, [2022] | Series: Shark files - a Crabtree seedlings book | Includes index.
Identifiers: LCCN 2021018415 (print) | LCCN 2021018416 (ebook) | ISBN 9781427158307 (hardcover) | ISBN 9781427158314 (paperback) | ISBN 9781427158321 (ebook) | ISBN 9781427158338 (epub) | ISBN 9781427158345
Subjects: LCSH: Bull shark--Juvenile literature.
Classification: LCC QL638.95.C3 L865 2022 (print) | LCC QL638.95.C3 (ebook) | DDC 597.3/4--dc23
LC record available at https://lccn.loc.gov/2021018415
LC ebook record available at https://lccn.loc.gov/2021018416